CW01551271

Christmas Science

25 Activities for Christmas Science

Copyright @ Elemental Science, Inc.

ISBN#978-1-935614-70-8

Produced by: Elemental Science

support@elementalscience.com

Copyright Policy

All contents copyright © 2019 by Elemental Science. All rights reserved.

No part of this document or the related files may be reproduced or transmitted in any form, by any means (electronic, photocopying, recording, or otherwise) without the prior written permission of the author. The author does give permission to the original purchaser to photocopy all game materials for use with the students in their classroom.

LIMIT OF LIABILITY AND DISCLAIMER OF WARRANTY: The publisher has used its best efforts in preparing this book, and the information provided herein is provided "as is." Elemental Science makes no representation or warranties with respect to the accuracy or completeness of the contents of this book and specifically disclaims any implied warranties of merchantability or fitness for any particular purpose and shall in no event be liable for any loss of profit or any other commercial damage, including but not limited to special, incidental, consequential, or other damages.

TRADEMARKS: This book identifies product names and services known to be trademarks, registered trademarks, or service marks of their respective holders. They are used throughout this book in an editorial fashion only. In addition, terms suspected of being trademarks, registered trademarks, or service marks have been appropriately capitalized, although Elemental Science cannot attest to the accuracy of this information. Use of a term in this book should not be regarded as affecting the validity of any trademark, registered trademark, or service mark. Elemental Science is not associated with any product or vendor mentioned in this book.

Table of Contents

Introduction

Dear Reader,

Every year, at Christmas, we like to take a break from our normal plans to spend the weeks leading up to the holiday enjoying some Christmas-themed activities. My kids are bouncing off the walls with excitement, so this break gives me a chance to make learning super fun. We read lots of Christmas-themed books, learn about how other cultures celebrate the holiday, and we do Christmas-themed science activities!

In this book, I am sharing the twenty-five science activities and six nature study ideas that we pick and choose from during the holiday season. We typically do several of the holiday-science activities on Monday through Thursday of the week and one of the winter nature study ideas for Friday during the month of December. But however you choose to add in a bit of scientific sparkle to your holidays, there are thirty-one options for you to use on the following pages.

My hope is that the ideas in this book will help you add a bit of holiday magic to your science plans!

Happy holidays,
Paige Hudson

How to Handle Homeschool Science During the Holidays

Have you ever heard of Christmas school?

It's when you hit pause on your regularly scheduled plans and set aside some time for holiday-themed activities. It's a super fun way to embrace the chaos that typically surrounds this season.

In our own homeschool, we have looked at Christmas celebrations around the world, we have studied the Nutcracker Suite, we have looked at the different Christmas symbols, and more.

But one thing I have always found lacking in the pre-planned Christmas units is the science part of the plans. I'm guessing that I am not the only one out there who has run into this.

And I know that the solution is to usually drop science from your plans altogether - it's OK, I promise I won't throw eggs at you if you have done this in the past.

But let me encourage you this year to not drop science. Give the gift of wonder and amazement instead!

In that spirit, I wanted to share with you all a few options for how to handle homeschool science during the holidays.

Homeschool Science During the Holidays

Wondering how you can handle homeschool science in the midst of the holiday craziness? Here are a few options for you!

Option #1 - Add science to your holiday homeschool plans.

As I said before, most of our holiday-homeschool units have lacked science, so, as a solution, I just added them in!

Sometimes, I have just thrown in a weekly holiday- or winter-themed science project. You can see a few of our favorites in December's tip of the month.

Sometimes, I have actually taken the time to pick and choose science activities that relate to what we are studying. We have a great seasonal science Pinterest board that you can mine through for ideas:

🖰 https://www.pinterest.com/elementalscienc/seasonal-science-experiments/

Option #2 - Create a full holiday unit around science projects.

You guys know I love science! And I am all about letting science take the lead as the rest of your plans line up behind it, which is exactly what we did for our son's preschool.

So, why not do the same for Christmas?

Last year, we did this with three holiday-ish materials - snow, cookies, and evergreens. We started our week with one of these Christmas-themed science activities and then added in math, reading, art, history, and music that related to the science material for the week. It was a blast!

You can do the same this year, or you can choose to use these daily Advent science ideas from Inspiration Laboratories.

Either way, letting science take the lead for your holiday homeschool plans is a sure way to provide heaps of holiday cheer!

Wrapping it up

There are so many wonderful holiday-themed science projects that I know your students will love. I trust that these two options will help you add a bit of science to your holiday festivities!

More Holiday Science Tips

Tips for Homeschool Science Podcast

* How to easily add a sprinkle of holiday science cheer - I love Christmas – the lights, the cookies, the warm feelings, and the holiday science experiments. Come listen to episode 18 of the Tips for Homeschool Science Show where I share how we add a sprinkle of Christmas science to your holiday season!
  ⌐ https://elementalscience.com/blogs/podcast/18

Homeschool Science Tip of the Month

* Christmas Science {December 2019} - Christmas science is perfect for the whole family, which is why we encourage you to let your older students in on the fun in this tip of the month!
  ⌐ https://elementalscience.com/blogs/homeschool-science-tips/christmas-science
* Science Gifts {December 2018} - December is here! Get a few science gift ideas in this homeschool science tip of the month.
  ⌐ https://elementalscience.com/blogs/homeschool-science-tips/science-gifts
* Holiday Science {December 2017} - Sprinkle some holiday cheer into your homeschool science plans with the information in this tip!
  ⌐ https://elementalscience.com/blogs/homeschool-science-tips/holiday-science
* Take a Break {December 2016} - I know I am not alone in turning the most mundane experience into a homeschool lesson, but everybody needs a break and that's what this tip is all about!
  ⌐ https://elementalscience.com/blogs/homeschool-science-tips/take-a-break

15 Minutes at the Window {Winter Nature Study}

Let's face it – during the winter months, our kids are not usually clamoring to head outside and run around in nature.

And if we are really honest, neither are we.

Though I do see value in heading outside to study nature up close during the winter months, it can be downright cold!

Enter fifteen minutes at the window.

This unpretentious form of indoor nature study is as simple as:

1. Sitting at a window that faces the outdoors.
2. Setting the timer for 15 minutes.
3. And watching what happens.

At first glance, it might look barren and devoid of nature, but I urge you to take another look.

You can study the lichens you see on the trees.

You can look for rocks peeking out of the soil.

You can discuss the falling snow (or the snow drifts, if you have a few.)

You can talk about why trees lose their leaves and how those leaves are decomposed.

You can sketch the difference between coniferous and deciduous trees in the winter.

You can watch for signs of the different kinds of wildlife.

If you really want to attract birds and other small animals, try setting out some food – such as bird seed, peanuts, or a slice of orange.

Not only will you get to look at your visitors up close, but you can share about the birds without fearing that your words will scare them away. While the animal is there, you can also discuss the adaptations the creature has to survive the cold temperatures.

Fifteen minutes at the window won't be exactly the same as heading outside and examining nature up close and personal. But, I trust that you can see the value in pausing to examine the life you can find outdoors from the comforts of your own home.

Science Activities

Candy Cane Magic

My kids prefer to make candy canes disappear into their stomachs, but they love to watch this magical experiment year after year!

Supplies Needed

To make some candy canes disappear, you will need the following:

* ❄ 2 Clear cups
* ❄ 2 Small candy canes (or 1 big candy cane broken into two pieces)

Procedure

1. Fill the first cup with cold water and the second with warm water.
2. Then, drop the candy canes into the cups at the same time.
3. Observe what happens.

Explanation

Both candy canes will dissolve in the water, but the one in the warm water will disappear much more quickly. It may even twist and change shape as it dissolves. This is because the heat speeds up the movement of the molecules in the water, causing the reaction to speed up!

Take it Further

Make some peppermint syrup to add to your hot chocolate or coffee! Add ½ cup of sugar, ½ cup of broken candy canes, and 1 cup of water to a pot. Heat the pot on the stove until the sugar and peppermints completely dissolve. Remove from the heat - you now have a delicious Christmas treat!

Christmas Maze

LEGOs are perfect materials to create a Christmas-themed maze. Let your kids' imaginations run wild with this simple activity - you'll be amazed by what they create!

Supplies Needed

To create your own Christmas maze, you will need the following:

* Flat LEGO board
* Variety of LEGO bricks in holiday colors
* Marble

Procedure

1. Lay out the flat LEGO board and explain to the students that they are going to create a maze for the marble to go through using the bricks. It needs to have a start and a finish, but other than that, the maze design is up to them!
2. Then, let them begin to design, build, and test their maze.

Explanation

The activity is simply a fun way for your students to practice their engineering skills!

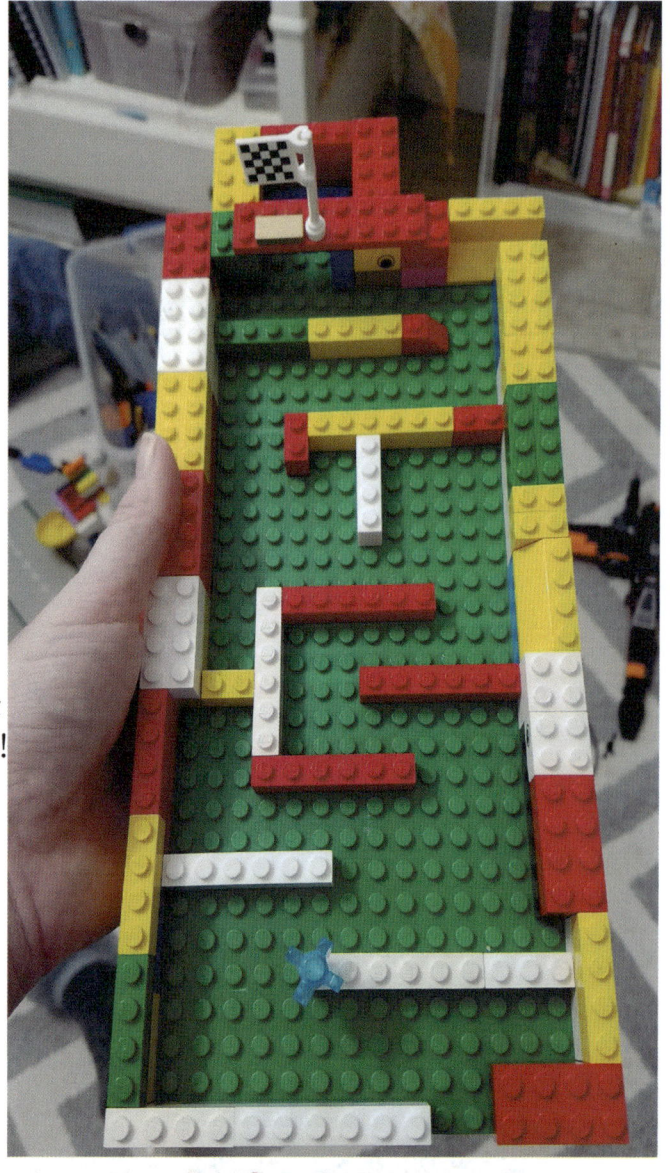

Take it Further

Have the students design another LEGO creation for Christmas!

Christmas Milk Magic

This is a Christmas twist on a classic experiment - magic milk!

Supplies Needed

To mix up a batch of Christmas milk, you will need the following:

* Warm milk
* Red and green food coloring
* Dish soap
* Shallow dish

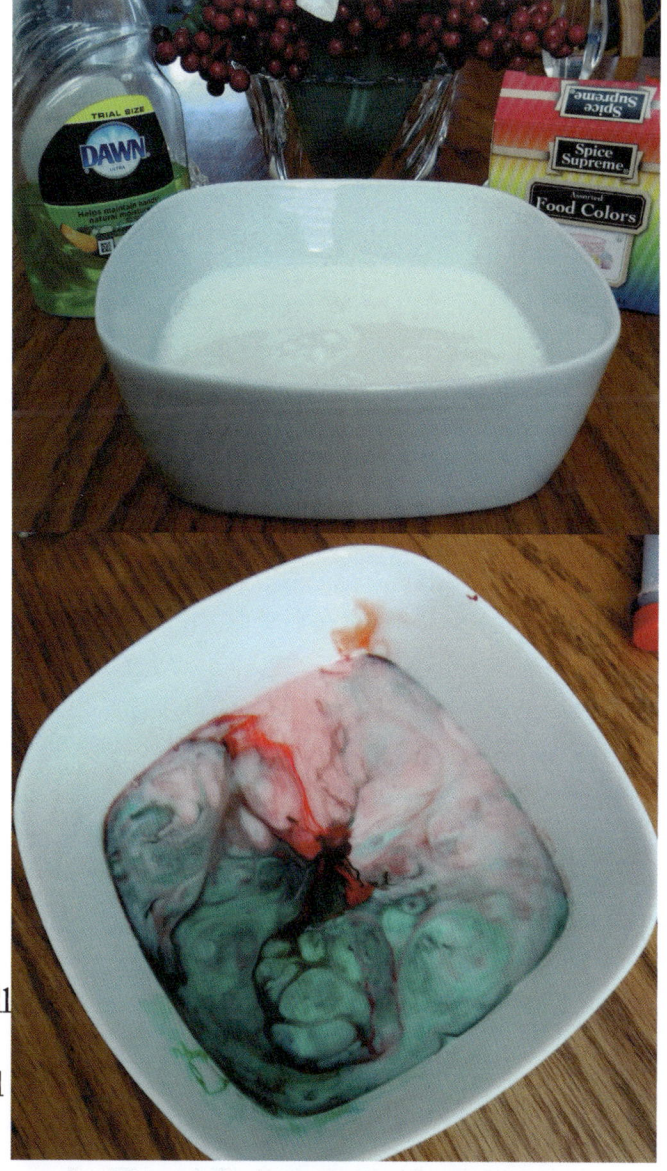

Procedure

1. Pour a thin layer of warm milk into a bowl.
2. Next, add a few drops of red food coloring to one end of the bowl and green to the other.
3. Have the students add a drop of liquid dish soap to the center of the bowl
4. Sit back and observe the colors as they set out on a swirling journey to make some Christmas-colored milk!

Explanation

The dish soap breaks the surface tension on the milk, causing the milk molecules to tumble and mix all around the bowl. As the molecules mix, they pick up the food coloring molecules, creating swirls of colorful liquid in motion throughout the bowl.

Take it Further

Have the students test to see if the results are different with ice cold milk.

Christmas Slime

Slime is always a big hit in our house, so as we get closer to Christmas Day, we like to whip up a festive version of classic Borax slime!

Supplies Needed

To whip up a batch of Christmas slime, you will need the following:

- ✻ Clear gel glue
- ✻ Water
- ✻ Glitter
- ✻ Green food coloring
- ✻ Borax Laundry Booster

Procedure

1. Begin by mixing 4oz of glue with 4oz of water, a few drops of green food coloring, and a shake of glitter in a plastic bag.
2. Next, in a separate cup, mix a quarter cup of water with half a teaspoon of Borax.
3. Then, add the Borax solution to the baggie and massage the bag for a few minutes until a nice firm slime has formed.
4. Pull the slime out of the baggie and have fun!

Explanation

Borax slime is the result of a reaction between the glue and the borax that forms an elastic polymer.

Take it Further

Have the students mix up another batch of slime in a different color. Then, see if they can get the two to mix.

Christmas Tree Cookies

You can't eat these cookies, but they still make a delicious scientific treat!

Supplies Needed

To make your own Christmas science cookies, you will need the following:

* Baking soda
* Vinegar
* Green food coloring
* Eye dropper
* A plate and a bowl
* A Christmas tree cookie cutter

Procedure

1. Mix several drops of the green food coloring with the vinegar in the bowl.
2. Then, place the cookie cutter on the plate and fill it with baking soda.
3. Now, use the eyedropper to squirt colored-vinegar on it!

Explanation

You should see lots of fizzing where the vinegar meets the baking soda. This is due to the gas, carbon dioxide, that is released when an acid, like vinegar, reacts with a base, like baking soda.

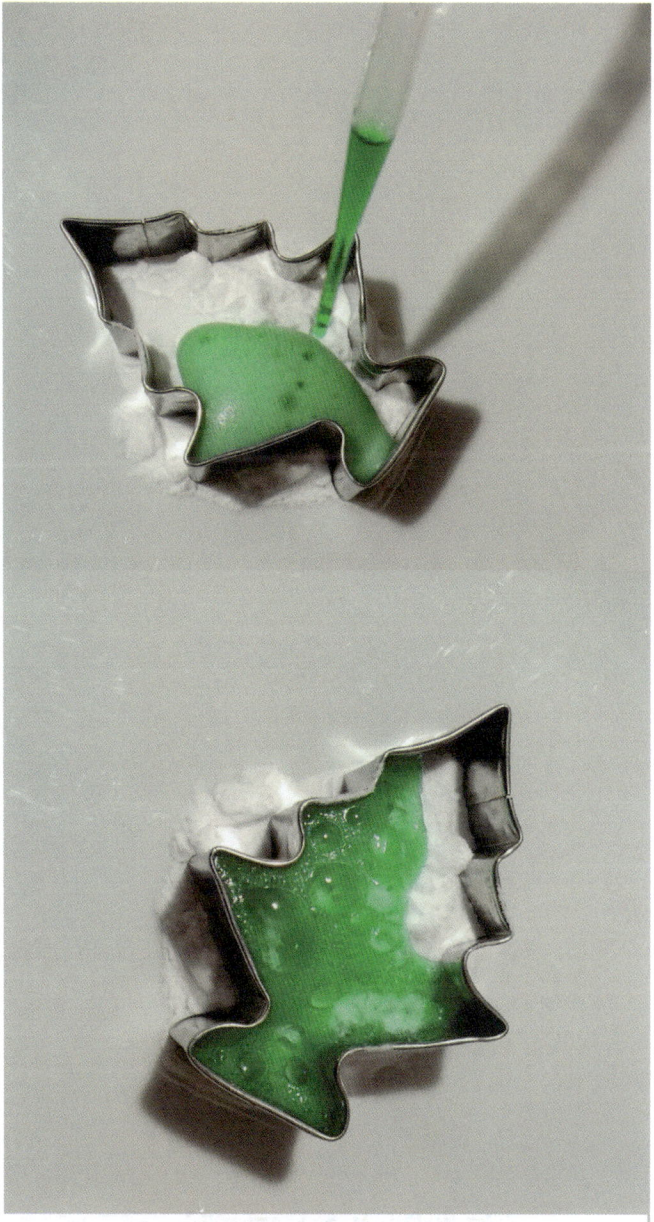

Take it Further

Have the students repeat this experiment with different shaped cookie cutters or with differently colored vinegar.

Chromatography Ornament

Supplies Needed

To make your own chromatography ornaments, you will need the following:

* ❋ Coffee filters and Rubbing alcohol
* ❋ A wide-mouthed jar or bowl
* ❋ Permanent markers in a variety of colors
* ❋ Laminating sheet or contact paper (optional)
* ❋ String and a hole punch

Procedure

1. Have the students create a design with different colored dots in the center of the absorbent material you have chosen. (*Note – Their designs will spread out quite a bit.*)
2. Next, pour a thin layer of rubbing alcohol into the bottom of a jar.
3. Then, fold the coffee filter in quarters and place the tip into the rubbing alcohol. Watch the design spread out. Remove the filter when it reaches the desired effect.
4. Let their creations dry and cut out the design, Laminate the ornament if you desire, punch a hole, tie the string through and hang it on the tree!

Explanation

Markers are made up of several different types of ink, which is soluble in isopropyl alcohol. The alcohol picks up the molecules of ink and carries them along the filter paper. Some of these ink molecules are heavier than the others and this technique separates the ink into different colors by depositing the heavier molecules sooner than the lighter ones.

Take it Further

Repeat over and over until your entire tree is decorated with science!!

Crystal Christmas Tree

Supplies Needed

To grow your own crystal trees, you will need the following:

* ✳ Cereal box cardboard
* ✳ Green food coloring
* ✳ Shallow dish or plastic bowl
* ✳ Water, Liquid bluing, Salt

Procedure

1. Use the Christmas Tree Templates on pg. 28 to cut out two trees from the cereal box cardboard. Cut out the black rectangles and fit the two together to form a 3-D tree. Then, add a bit of green food coloring to the edges of the cardboard tree and place the tree in the shallow dish.
2. Next, mix together 2 TBSP each of water, salt, and liquid bluing and pour the mixture into the shallow dish. Set the dish where it won't be disturbed, but will still have good air flow.
3. The next day, sprinkle in two more tablespoons of salt.
4. On day three, pour 2 TBSP each of water, salt, and liquid bluing into the dish, but not directly over the cardboard tree. At this point, you should be seeing crystals forming, but if not, then add 2 TBSP of ammonia to the bowl.
5. Let your crystal Christmas tree keep growing until you reach the desired effect!

Take it Further

Have your students repeat this process using several different Christmas-themed shapes!

Crystal Ornament

Supplies Needed

To build your own festive toothpick-gumdrop castle, you will need the following:

* A wide mouthed jar
* A pencil
* Several pipe cleaners
* Borax Laundry Booster

Procedure

1. Start by shaping pipe cleaners to into a snowflake shape. This can be as simple or as complex as you wish, but make sure it will fit through the opening of your jar.
2. Next, attach your snowflake to the pencil. You want the pencil to be able to rest on the edge of your jar without having your snowflake touch the sides or bottom of the jar.
3. Now, add hot water until it almost fills the jar, noting how many cups of water it takes to fill the jar.
4. Then, add the Borax, one tablespoon at a time, taking care each time to stir until the Borax is dissolved. You want to add about 3 tablespoons of Borax for every cup of water you have added.
5. Finally, hang your snowflake in the jar so that it is completely covered by the liquid. Allow the jar to sit undisturbed overnight.

Explanation

The next morning, you should see your design covered in beautiful crystals. These were formed as the Borax came out of solution and attached itself to the pipe cleaners.

Take it Further

Have the students make several other different shapes to place in the Borax solution to make ornaments. Then, use tomorrow's activity to decorate an entire tree with science!

Disappearing Snow Magic

Do you know how much water is in snow? The cool winter months are the perfect time to have a race to find out where the line will end up as the snow melts!

Supplies Needed

To have a snowmelt race, you will need the following:

* A tall plastic container
* Snow (or ice, if you don't have any available)
* Masking tape
* Pen

Procedure

1. Fill the tall plastic container with snow (or crushed ice cubes, if you don't have any snow on the ground).
2. Have the students guess where the line will be when the snow (or ice) melts, and use the masking tape to mark the spots.
3. Check the progress every hour or so, until the snow or ice completely melts.

Explanation

Water is one of those molecules that occupies more space as a solid than it does as a liquid. So, as the snow (or ice) melts, the amount of space in the jar occupied by the water will decrease. In the end, the liquid water will be about a quarter of what it original solid water was.

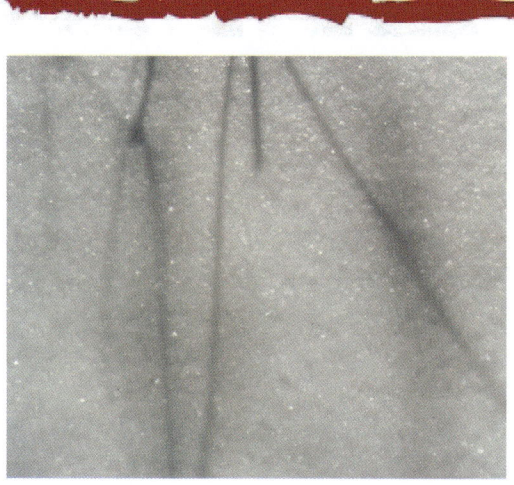

Take it Further

Have the students use a magnifying glass or palm-sized microscope to observe the snow up close.

Gingerbread Engineering

I don't remember making a Gingerbread House as a kid. So, when I first made one with our kids, I had no idea how planning and engineering went into the process!

Supplies Needed

To engineer your own Gingerbread House, you will need the following:

* A Gingerbread House Kit
* A batch of royal icing
* Cardboard

Procedure

1. Lay the pieces out from your gingerbread kit and plan how to put your house together.
2. Use the icing to adhere the pieces.
3. As you build, talk about how you need to support the pieces so that the house will stand on its own. Also discuss how the amount of icing affects the building process. Make any necessary adjustments along the way, based on what the students have learned.

Explanation

An engineer has a problem, he imagines and designs a solution, he tests that solution, he makes any adjustments, and he completes the project. As your students build their gingerbread house, they are using the same basic steps that an engineer uses!

Take it Further

Have the students build a gingerbread castle out of Legos!

Gumdrop Trees

At Christmas time, we like to give a classic gumdrop engineering experiment a holiday twist!

Supplies Needed

To build your own festive toothpick-gumdrop tree, you will need the following:

* A box of toothpicks
* A plate and a bowl
* Red, green, and white gumdrops

Procedure

1. Set out the gumdrops in a bowl and the box of toothpicks on the table.
2. Give each student a plate to build on. (*This is not absolutely necessary, but it does make cleanup a lot easier!*)
3. Let your students design, test, and build whatever gumdrop tree they can dream up.

Explanation

This activity is a fun way for your kids to test the strength of different shapes. They should see that certain shapes, like the cube, are stronger than others. They should also see that there is a limit to how much height a certain-sized base can hold before their tower will collapse on itself.

Take it Further

Test whose structure can hold the most weight! To do this, you will need a small plastic container and lots of pennies. Set the container on top of a gumdrop castle and add pennies until the structure collapses. Then, count the number of pennies the gumdrop castle could hold!

Holiday Fireworks

Who doesn't love fireworks? But the mini-explosions aren't so safe to do in the kitchen! However, these holiday fireworks will allow you to have a colorful explosion without blowing anything up!

Supplies Needed

To have some holiday fireworks, you will need the following:

* Clear glass jar
* Water
* Oil (vegetable)
* Green and red food coloring

Procedure

1. Pour a bit of oil in a bowl and add a few drops of food coloring.
2. Break the drops into tiny droplets with a fork.
3. Slowly add the oil mixture to a jar filled part way with water.
4. Wait a moment and observe your fireworks!

Explanation

Oil and water don't mix. This is because one (oil) is nonpolar and the other (water) is polar. Polar molecules prefer to stay with other polar molecules and the same goes for nonpolar ones. Food coloring is also polar, just like water, so it doesn't want to mix with the oil either. When gravity takes effect, the food coloring droplets fall from the oil layer into the water layer.

Take it Further

Have the students repeat the experiment with different temperatures of water to see how the results differ.

Holiday Spinners

These super-fun holiday toys are made from simple materials, plus your students will learn about the physics of spin and motion as they play!

Supplies Needed

To make your own holiday spinners, you will need the following:

* CD
* Marble
* Plastic water bottle cap
* Hot glue
* Permanent markers (in a variety of colors)

Procedure

1. Have the students use the permanent markers to decorate their CDs with a holiday design.
2. (**Adults Only**) Use the hot glue to attach the bottle cap to the center of the decorated side of the CD. Then, flip the CD over and use the hot glue gun to attach the marble to the center of the CD.
3. Once the glue dries, let the kids play with their holiday spinners by gripping the bottle cap and giving it a spin.

Explanation

The marble reduces the friction, allowing the CD to spin when your student applies a bit of force to the cap.

Take it Further

Have the students test different amounts of force applied to the bottle cap to see how it affects the motion of the holiday spinner.

Ice Crystal Paintings

These paintings are not really made from ice crystals, but the end result looks just like ice crystals!

Supplies Needed

To create your own ice crystal paintings, you will need the following:

* Epsom salts
* Hot water
* Plastic cup
* Food coloring
* Construction paper

Procedure

1. Mix equal parts of Epsom salts and hot water together until most of the Epsom salts have dissolved.
2. Add a few drops of food coloring and mix well. Then, set this mixture out on the table.
3. Have the students use the mixture to paint a snowflake design or ice storm on the construction paper.
4. As it dries, the ice crystals will form!

Explanation

In step 1, you made a super-saturated solution of Epsom salts for the paint. As the paint dried, the water evaporated and the Epsom salts came out of solution to form crystal solids once again.

Take it Further

Have the students set the cup with the remaining Epsom salt paint out on a counter where it won't be disturbed. Check the cup daily and observe the changes. (*After about a week, you should see crystals form in the cup.*)

Indoor Snow

If you don't have snow in your area, this experiment is perfect - it's super easy and even feels a bit cool to the touch!

Supplies Needed

To mix up a batch of indoor snow, you will need the following:

* ❋ Box of cornstarch
* ❋ Can of Shaving cream
* ❋ A pie plate

Procedure

1. Add half a box of cornstarch to the pie plate.
2. Squirt in about half a can of shaving cream and mix till you achieve a snow-like substance that can be molded!
3. Then, let your students play until their hearts are content.

Explanation

The reaction that occurs when you mix the cornstarch and shaving cream is an endothermic one, which means that it takes in heat, making the resulting product feel cool to the touch!

Take it Further

Have the students build a snowman, either out of indoor snow or out of the real stuff you can find outside!

Marshmallow Catapult

With this activity, you will build your own catapult to launch marshmallows into a cup of hot cocoa!

Supplies Needed

To make your own marshmallow catapult, you will need the following:

* 8 Jumbo (wide) popsicle sticks
* 3 Rubber bands
* Plastic bottlecap, Glue
* Mini-marshmallows

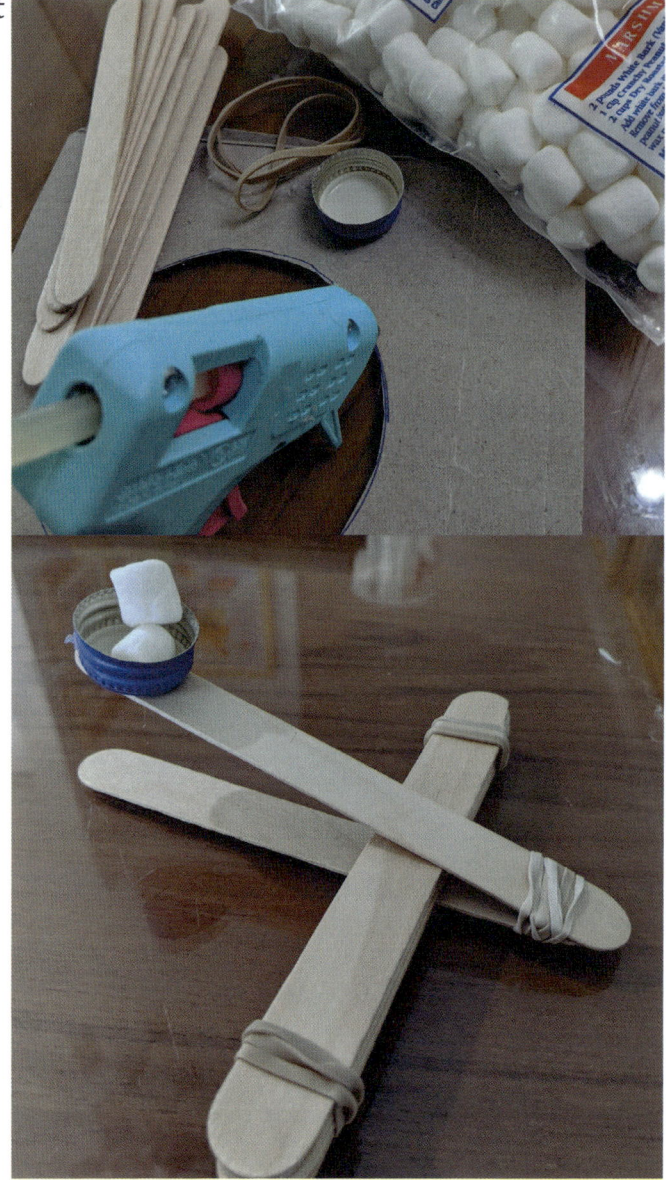

Procedure

1. Begin by gluing the bottle cap to the top of one of the popsicle sticks to make a cup for the marshmallows.
2. Next, take 6 popsicle sticks, stack them on top of each other. Use two rubber bands to tie them together at each end.
3. Glue the remaining popsicle stick about one third of the way down the stick, halfway in between the bottom of the stack.
4. Flip the stack over and use the final rubber band to attach the popsicle stick with the bottlecap to the single stick. (*See picture for reference.*)
5. Add a marshmallow to the bottlecap. Gently push the stick with the bottlecap down and then release!

Take it Further

Have the students have a contest to see who can get the most marshmallows into a cup of hot cocoa!

Nature Viewer Ornament

These ornaments are a wonderful way to enjoy the beauty of the outdoors from the comforts of your living room couch!

Supplies Needed

To make your nature viewers, you will need:

* Objects from nature
* Clear, plastic ball ornaments with removable tops

Procedure

1. Head outside to collect objects from nature that will fit into the hole at the top of your ornament - things like air plants, berries, small nuts, leaves, or lichen.
2. Open up the top of the clear, plastic ball ornament and add the objects that were collected. You can shake or swirl the ornament around a bit to get the nature objects into the places you want.
3. Then, carefully put the tops back on and hang the ornaments on the tree.

Take it Further

Use some of the sticks and logs you find as you search for objects in nature to make an old-fashioned sawhorse style reindeer for your yard. You will need two smallish logs for the body and head, five medium-sized sticks for the neck and legs, and three small branched twigs for the tail and antlers. Attach the sticks and twigs to the logs in the shape a reindeer. Then, decorate your reindeer with bows and a red nose!

Newspaper Tree

Have your students learn about physics and practice engineering by creating a newspaper tree!

Supplies Needed

To make your own newspaper tree, you will need the following:

* Newspaper
* Tape
* Ornaments or various weights

Procedure

1. Give the students a stack of newspaper and some tape. Explain to them that they are going to use the newspaper and tape to design and build a tree that will hold ornaments.
2. Help them come up with a plan (e.g., you can use several diminishing triangles of rolled newspaper to create a tree or you can create a central column with rolled newspaper branches).
3. Once they have a design, let them build their newspaper trees.
4. Then, hang several ornaments on the tree to see which design can hold the most ornaments!

Explanation

The activity is simply a fun way for your students to practice their engineering skills!

Take it Further

Have the students use the principles they learned to create another, stronger newspaper tree!

Ornament Physics

So, at this point in our Christmas science journey, we have created four unique ornaments. Now, it's time to use a bit of physics as we hang these on the tree!

Supplies Needed

To learn about physics as you hang your ornaments, you will need the following:

* Ornaments of varying weights
* A Christmas tree

Procedure

1. Choose several ornaments, one light, one heavy, and one in between.
2. Next, select a branch to hang them on.
3. Hang the first ornament and observe what happens to the branch.
4. Repeat Step 3 with the remaining ornaments you selected. Observe any changes to the branch each time.

Explanation

The students should see that the heavier the ornament, the more the branch bends under the weight. This is because the heavier the ornament, the more downward force it exerts on the branch.

Take it Further

Have the students repeat the experiment with different thickness of branches to see how this changes their results. (*The thicker the branch, the less it will bend, as the branch is able to handle more weight without changing.*)

Peppermint Bows

Can you tie a candy cane in a bow? With a little heat you can!

Supplies Needed

To build your own peppermint bows, you will need the following:

* Several candy canes
* Wax paper
* Cookie sheet
* Oven

Procedure

1. Preheat the oven to 250°F.
2. Line a cookie sheet with wax paper.
3. Place the candy canes on the cookie sheet.
4. Put them in the oven for 3 to 4 minutes.
5. Use oven mitts to remove the candy canes and then gently twist and bend them into any shape you like!
6. Allow your creations to cool and observe!

Explanation

Candy canes are made from mostly sugar crystals that are held together, forming the candy cane shape. When you heat the candy cane up, the bonds holding the crystals in place weaken, allowing you to twist the candy into a different shape. As it cools, the bonds reform and the structure stays in its new shape.

Take it Further

Have the students try to twist candy canes that haven't been warmed up - is there a difference?

Rudolph's Toothpaste

This is another Christmas twist on a classic experiment - elephant toothpaste!

Supplies Needed

To mix up a batch of reindeer toothpaste, you will need the following:

* Yeast, Water, Cup
* Empty water bottle, Hydrogen peroxide, Red food coloring, Liquid dish soap

Procedure

1. In the cup, mix together a packet (2 tsp) of yeast with a quarter of a cup of warm water and set aside.
2. In the bottle, mix half a cup of hydrogen peroxide, 3 to 5 drops of red food coloring, and 7 to 10 drops of liquid dish detergent.
3. Once the yeast blooms, add it to the mixture in the bottle and watch what happens!

Explanation

The students should see that the mixture bubbles up, creating a foam that pushes up and out of the bottle. Hydrogen peroxide breaks down into water and oxygen naturally, but the reaction is quite slow. An enzyme in yeast, catalase, speeds up this reaction. The dish soap traps the oxygen bubbles, creating the foam that is pushed up and out of the bottle!

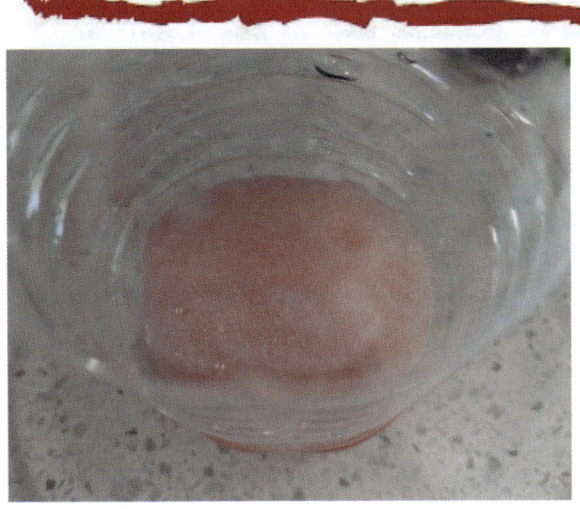

Take it Further

Have the students repeat the experiment, only this time do not add dish detergent. Observe how this changes the results.

Salty Ice

Supplies Needed

To see how salt affects the freezing point of water, you will need the following:

* 3 Cups
* Water
* Food coloring
* Salt

Procedure

1. Begin by adding one cup of water to each of the cups. Label them as #1, #2, and #3.
2. Then, add several drops of food coloring to cup #2 and add 3 TBSP of salt to cup #3. Mix both cups well.
3. Next, place all three of the cups in the freezer.
4. Check the cups every 30 minutes for 3 hours to observe what is happening.

Explanation

The students should see that the cup #1 and cup #2 freeze at the same time, while cup #3 takes quite a bit longer to freeze. This is because salt lowers the freezing point of water, which means that water with salt in it will remain a liquid for longer than plain water because the point at which salt water will freeze is lower than 32°F. Food coloring has no effect on the freezing temperature of water, so the colored water will freeze at the same temperature as the plain water.

Take it Further

Have the students try this experiment outdoors! Find a patch of ice and sprinkle some Ice Melt (rock salt) on it. Then, check it every so often to observe what happens.

Simple Circuit Ornament

Light up your tree and learn about electricity with this simple circuit ornament!

Supplies Needed

To make your own simple circuit ornaments, you will need the following:

* Cereal box cardboard
* LED light bulb
* Watch battery
* Tape
* Markers and/or glitter

Procedure

1. Cut out a simple ornament shape from the cardboard using the template on pg. 27 of this guide.
2. Cut a small hole in the center of the ornament large enough for for the light bulb to fit through.
3. Have the students decorate the ornament with the markers and glitter. (*You can also cut out and decorate another ornament shape for the back if you desire.*)
4. Once their decorations are dry, insert the light bulb through the slit and then place the battery in between the wires so that the light bulb turns on.
5. Use the tape to secure the battery in place and then hang the ornament on your tree!

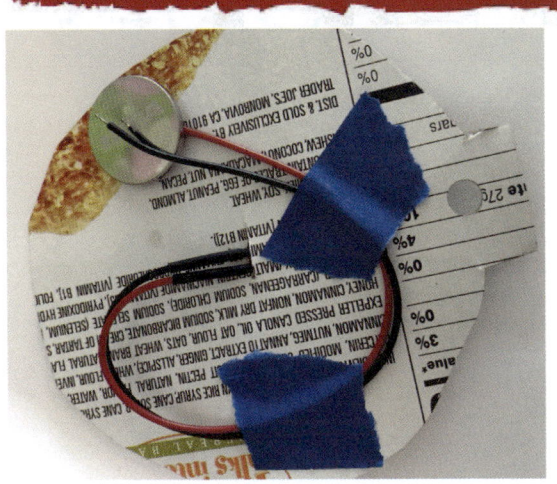

Take it Further

Have the students use the same principles to make a Christmas card for someone in their family!

Underwater Candle Magic

Can you make a candle burn underwater? With a bit of science magic, you can!

Supplies Needed

To do this magic trick, you will need the following:

* Long taper candle
* Air dry clay
* Glass or jar
* Ice water
* Matches

Procedure

1. Cut the candle so that the top of it just reaches the top of your glass. Then, use the air dry clay to fix the candle to the center of the bottom of the glass.
2. Add ice water so that the level is just below the wick of the candle.
3. (**Adults Only**) Light the candle and watch what happens! (*Note – To extend the length of time the candle burns underwater: Every 5 to 7 minutes, siphon off a bit of the water with an eyedropper and replace the same amount with fresh ice water.*)

Explanation

The ice water cools and solidifies the wax, creating a tunnel that the candle can continue to burn down into, despite the fact that the flame is under the water level.

Take it Further

Have the students repeat the experiment with different sized candles to see how the results differ.

Woolen Mittens

Do mitten really help your hands stay warm? You'll find out with this activity!

Supplies Needed

To test your mittens, you will need the following:

* Two small balloons
* Warm water
* A woolen mitten or hat

Procedure

1. Fill the two balloons with warm water.
2. Stuff one of the balloons into the mitten and leave the other one unprotected.
3. Set both balloons outside on a cold day. (Note - If you don't have a cold day, use the refrigerator!)
4. After 10 minutes, go feel each balloon to see how warm they are. Keep repeating this process until both balloons are cold.

Explanation

The students should that the balloon inside the mitten stayed much warmer than the one exposed to the elements. This is because the materials used to make mittens trap heat, preventing the item inside from cooling off to quickly. Unlike our hands, the balloon doesn't make its own heat, so eventually it does cool off.

Take it Further

Have the students repeat this experiment with different types of mittens to determine which one works the best.

Winter Nature Study

Cardinal

The male cardinal is an easy bird to spot, thanks to its bright red plumage. The cardinal is one of the first birds our daughter learned to spot and it remains one of her favorites!

Cardinals have a rare reddish beak, which is offset by a circle of black feathers close to their eyes. The male is dressed from head to toe in beautiful red feathers, while the female has a crown of red with a dress of greenish-gray feathers, tipped in red.

Cardinals are typically found near bushes where they like to build their nests. The females typically lay around 3 to 4 eggs, which they sit on till they hatch. The male will attend to the female, bringing her food, as she waits for the eggs to hatch.

The young of the cardinal are dull colored and have darker bills. This is so that they don't stand out, which would make the little fledglings easy targets for predators. The families will stay together throughout the winter, and then the fledglings will head out the following spring to find mates of their own.

Fun Fact - The cardinal is the state bird for seven different states!

Coordinating Science Activity

Listen for Cardinals – Head outside and listen for the call of the cardinal. (*Note - You can find this through the Audubon bird app: http://www.audubon.org/apps.*)

Christmas Lichen

The first time I saw Christmas lichen I was stunned. This dinner-plate-like lichen is festively colored and perfect to learn about during December, even though it only grows in along the Gulf Coast and coastal plains.

The Christmas lichen is a brightly colored example of a crustose lichen. These types of lichens are often round, flat, and somewhat crusty. Crustose lichens typically grow on tree trunks, but you can also find them on the flat surfaces or rocks as well.

Lichens are an example of a mutual beneficial biological relationship, typically between a fungus and an alga or a cyanobacterium. The alga or bacterium provides the fungal partner with sugars from photosynthesis, while the fungus provides protection from the environment for the alga or bacterium.

In the case of the Christmas lichen, the red hue comes from a chemical produced by one of the partners that make up this type of lichen. This red color typically collects around the outside and towards the middle. Scientists aren't exactly sure why yet, but doesn't it create the most amazing natural Christmas wreath?

Fun Fact - The Christmas lichen is actually used as a natural dye in Brazil.

Coordinating Science Activity

Lichen Hunt – Take the students on a nature walk to look for different kinds of lichens and where they grow. When you find a lichen, use a transparency (or tracing paper) to cover and trace it.

Holly

Holly adds lots of Christmas color and cheer to this time of year, which makes it a perfect subject to look at during the winter months. We are lucky enough to have a holly in our front yard, so we spent a few moments outside studying this plant before we ran back inside to warm up with a cup of hot cocoa!

Most hollies are evergreen, so they keep their leaves throughout the winter months. Their foliage is easily identified because of their dark green and glossy color, although some varieties have a touch of white. The holly leaf also typically has at least one spine, but can have up to fifteen!

Holly blooms in May or in June. Once pollinated, the female holly plant will form tiny ball-shaped fruits that mature around October. These berries are commonly thought of as red, but they can also be orange, white, or blue! Although the holly fruit is an important source of nutrients for birds during the winter months, it is inedible for humans.

Hollies are typically trimmed into bushes or hedges, but these plants can grow into trees that are forty to fifty feet in height!

Fun Fact - The holly bush was first used as a winter decoration by the Celts who used it to decorate their homes when celebrating the Yule festivities.

Coordinating Science Activity

Identify your Hollies – Head outside to find a few different types of hollies. Have the students make several observations as they record the number of spines on the leaf and the color of the berries. Then, with the assistance of a field guide, have them use their observations to determine the species of holly they have found.

Pine Tree

Every year, many of us welcome pine trees into our homes in the form of Christmas trees. These ornamented trees have been a part of holiday traditions all over the world, and the tradition dates back from the sixteenth century.

Pine trees are one of the oldest types of trees in the world – they live long, to the age of one hundred years old or more! There are around thirty-five different types of these evergreen coniferous trees growing in North America.

Pines are called evergreens because their needles (or "leaves") stay green all year long. The needles can last for up to two years, but when the old needles fall, new ones quickly replace them. There are a variety of ways to identify different types of pine trees. One way is by determining the number of pine needles there are per bundle (formally known as fascicle).

Pines are part of a group of trees known as coniferous, because they produce cones that contain seeds. These trees, like pines, spruces, and redwoods, are classified as gymnosperms because they reproduce by means of an exposed seed instead of one that is encased in a fleshy fruit.

Fun Fact - The Bristlecone Pine is the species of pine trees that lives the longest; some live to be over 5000 years old!

Coordinating Science Activity

Identify your Pines – Head outside to find a few evergreen trees. Have the students make several observations as they record the needle length and how many needles are found in a bunch. Have them also make a rubbing of the bark and record some general observations about the texture and color. Then, with an app like VTreeID or a field guide, have them use their observations to determine the species of tree you found.

Nuts

It's that time of year again – the time when you are afraid to walk under any type of tree that has nuts! Gravity doesn't help the situation, but the squirrels assist the situation by hurling nuts towards the ground at lightening speeds, in an effort to break their tough exteriors and enjoy the yummy goodness inside.

True nuts are dried fruits with a single seed inside. The seed case wall of a nut becomes very hard at maturity, but inside, the seed is packed with oils to help fatten up animals for the winter months. Their hard case protects the soft seed, or meat, inside.

Some nuts, like pecans, grow on trees; others, like hazelnuts, grow on large bushes. Most nut trees and bushes take three years or longer to mature and begin to produce nuts.

There are lots of seeds that we call nuts that are not actually true nuts. Walnuts and cashews are drupes, with two seeds inside. Peanuts are actually legumes. Brazil nuts have multiple seeds encased in a capsule or pod that splits apart.

Fun Fact – Acorns are actually considered true nuts.

Coordinating Science Activity

Dissect a Nut – Head outside to look for nuts. Collect a few and once you are back home, choose one or more of the nuts to dissect. Begin by observing the outside of the nut, then remove any covering and crack the nut open. Have the students observe the inside of the nut. You can also open several nuts and compare the differences and similarities.

Snow

December is usually the time when Old Man Winter decides to show up and gives us a bit of snow to enjoy. While you are outside playing, take a few moments to learn about this weather phenomenon!

Snow forms by a process of deposition, which means that water vapor high in the atmosphere changes directly into ice without becoming a liquid first. The temperature must be below freezing (32°F) for this to occur. If snow meets any warm air as it falls to the ground, it can be turned into rain, sleet, or freezing rain.

Snowflakes come in many shapes and sizes, but each one is six-sided. They form from as many as two hundred ice crystals, which come together in a lattice structure around a tiny piece of dust or dirt.

Fun Fact – Snow is white because the crystalline structure reflects all light, making it appear white to our eyes.

Coordinating Science Activity

Snow Engineering – Design and build a snow fort. You can dig it out or build it up. Either way, planning and creating a snow fort is a great way to learn about engineering!

Lists and Templates

Master Supply List

You will need the following supplies to complete all the activities in this book:

* A batch of royal icing
* A Christmas tree
* A Gingerbread House Kit
* Air dry clay*
* Baking soda*
* Borax Laundry Booster*
* Bowl
* Candy canes (at least 5)*
* CD*
* Cereal box cardboard*
* Christmas tree cookie cutter*
* Clear gel glue*
* Clear, plastic ball ornaments with removable tops*
* Coffee filters*
* Construction paper
* Cookie sheet
* Cornstarch (box)*
* Dish soap
* Epsom salts*
* Eye dropper*
* Flat LEGO board*
* Food coloring*
* Glass
* Glitter*
* Glue*
* Gumdrops (red, green, and white)*
* Hot glue
* Hydrogen peroxide
* Jumbo (wide) popsicle sticks (8)*
* Laminating sheet or contact paper (optional)
* LED light bulb*
* Liquid bluing*
* Long taper candle*
* Marble*
* Masking tape*
* Matches*
* Milk
* Mini-marshmallows
* Mitten
* Newspaper
* Objects from nature
* Oil (vegetable)
* Ornaments of varying weights
* Oven
* Pen*
* Pencil*
* Permanent markers in a variety of colors*
* Pie plate
* Plastic cups (7)*
* Plastic water bottle cap (2)*
* Plastic water bottle (empty)
* Plate
* Rubber bands (3)
* Rubbing alcohol
* Salt*
* Several pipe cleaners*
* Shallow dish
* Shaving cream (can)
* Small hanging ornaments
* Snow (or ice, if you don't have any available)
* String*
* Tall, plastic container*
* Tape*
* Toothpicks*
* Two small balloons*
* Variety of LEGO bricks in holiday colors
* Vinegar
* Warm water
* Watch battery *
* Wax paper*
* Wide-mouthed jar*
* Yeast*

Note - The items marked with a " * " are included in the Christmas Science experiment kit, which you can purchase from Elemental Science through the following link:

🖱 https://elementalscience.com/products/christmas-science-experiment-kit

Christmas Tree Template

1. Trace the shapes on cearal box cardboard.

2. Cut the trees out.

3. Cut a slit on the lines.

4. Slide the two pieces together on the slits to form a tree.

Ornament Template

1. Trace the shape on cearal box cardboard.

2. Cut it out.

3. Punch hole here

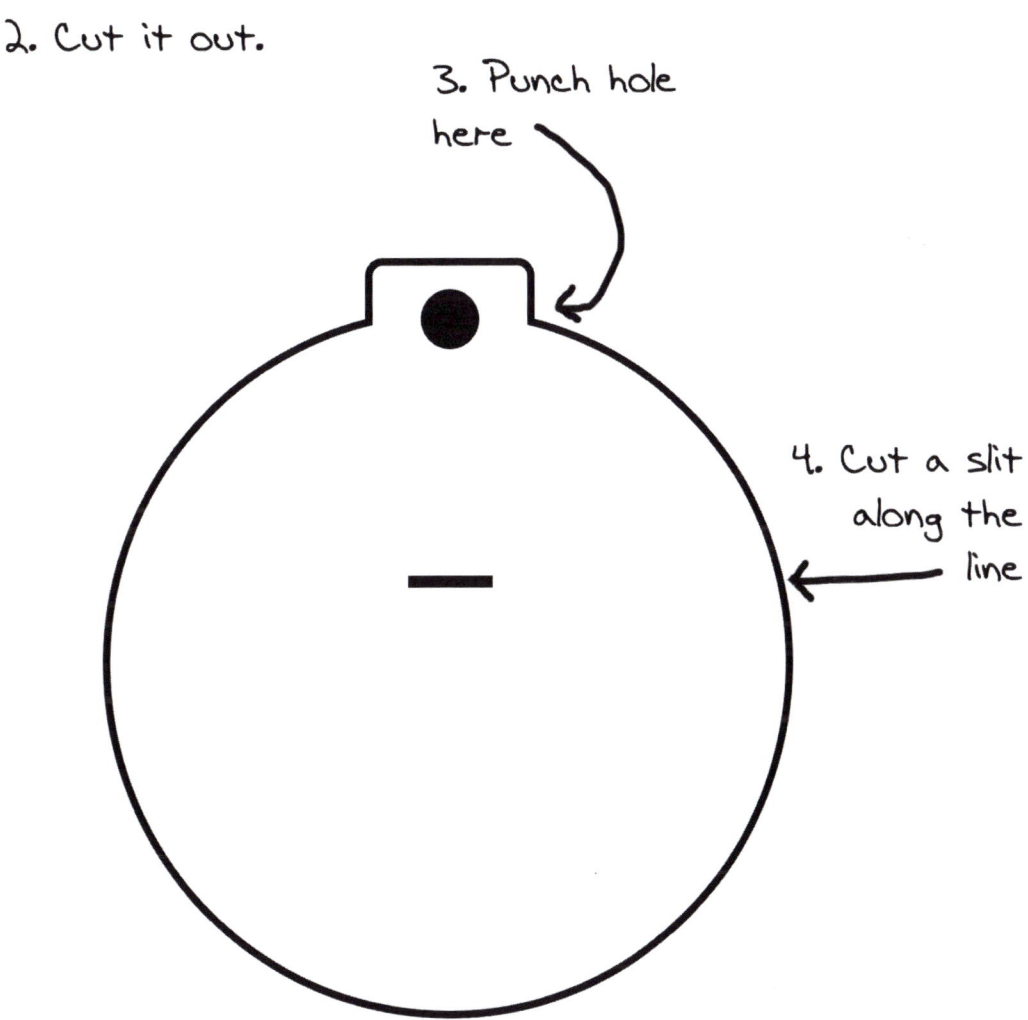

4. Cut a slit along the line

5. Repeat steps 1 through 3 if you want a back.

Printed in Great Britain
by Amazon

50418137R00030